TEC **ACPL ITEM** 2/02
DISCARDED

HOW CAN I EXPERIMENT WITH ... ?

A LEVER

David and Patricia Armentrout

Rourke
Publishing LLC
Vero Beach, Florida 32964

© 2003 Rourke Publishing LLC

All rights reserved. No part of this book may be reproduced or utilized in any form or by any means, electronic or mechanical including photocopying, recording, or by any information storage and retrieval system without permission in writing from the publisher.

www.rourkepublishing.com

PHOTO CREDITS: ©Armentrout Cover, pgs 7, 13, 14, 20; ©James P. Rowan pgs 9, 29; ©David French Photography pgs 17, 19, 23, 25, 27; ©Digital Vision Ltd. pg 4; ©Myrleen Ferguson Cate/PhotoEdit/PictureQuest pg 11.

Cover: *You use a lever when you remove a nail with the claw end of a hammer.*

Editor: Frank Sloan

Cover and page design: Nicola Stratford

Series Consulting Editor: Henry Rasof, a former editor with Franklin Watts, has edited many science books for children and young adults.

Library of Congress Cataloging-in-Publication Data

Armentrout, David, 1962-
  How can I experiment with simple machines? A lever / David and Patricia Armentrout.
      p. cm.
  Summary: Defines levers, explains their functions, and suggests simple experiments to demonstrate how they work.
  Includes bibliographical references and index.
    ISBN 1-58952-334-2
    1. Levers—Juvenile literature. [1. Levers--Experiments. 2. Experiments.] I. Title: Lever. II. Armentrout, Patricia, 1960- III. Title.
    TJ147 .A758 2002
    621.8—dc21

2002007651

Printed in the USA

w/w

Table of Contents

Making Work Easier	6
Simple and Complex Machines	8
Levers	10
Understanding Levers	12
Experiment with a Seesaw	15
Make a First-class Lever	16
Changing Force and Distance	18
Weight Arm and Force Arm	21
Make a Second-class Lever	22
Use Your Second-class Lever	24
Make a Third-class Lever	26
Common Levers	28
Glossary	30
Further Reading/Websites to Visit	31
Index	32

Lever (LEV er) — a rigid bar that pivots, or turns, on a fixed point; a simple machine used to move or lift objects

Scissors are double levers joined by a pivot point, or fulcrum.

Making Work Easier

Machines make work easier. Some machines, like cranes and bulldozers, tackle big jobs. They can lift and move heavy loads.

Think about life without these machines. How would we build our homes and roads? It would be hard to do these things without machines.

Even small machines make work easier. A wheelbarrow, for example, can't hold as much dirt as a bulldozer shovel, but it is still a machine that makes work easier.

A digger can easily move huge loads of dirt.

Simple and Complex Machines

Simple machines have few parts or no moving parts at all. The wheel, the pulley, the wedge, the screw, the inclined plane, and the lever are simple machines.

Complex machines are made up of simple machines. Complex machines have many parts. You can find complex machines around your house. Cars, air conditioners, and washing machines are machines made up of many parts.

A washing machine is a complex machine that makes daily chores easier.

Levers

Levers, like most other machines, give us a **mechanical advantage**. A mechanical advantage is what you gain when a simple machine allows you to use less effort. Levers reduce the amount of effort, or force, we use.

A lever, in its simplest form, is a board that rests on a balance point. If you picture a playground seesaw you can see a lever in its simplest form.

You could find plenty of levers around your house, if you knew what to look for. Can openers, nutcrackers, scissors, tweezers, pliers, and brooms are all types of levers. You'll understand how these objects are levers as you read on.

When you combine muscle power with a broom you put a lever in action.

Understanding Levers

Levers need three things to work. The first is the load. The load is the object, or weight, that needs to be moved. The second is the **fulcrum**. The fulcrum is the support, or balance point. The third is the effort. Effort is the force used to move the load.

Imagine you are given a task. You need to lift your friend off the ground. Your friend is smaller than you, but weighs more than you can comfortably lift. Suddenly your task is hard work! You need to find an easier way to get the work done. How about using a seesaw?

Lifting a friend off the ground is hard work.

Experiment with a Seesaw

Ask your friend to sit on one end of a seesaw. Now, you push down on the other end. That task wasn't so difficult! You were able to lift your friend easily. The seesaw gave you a mechanical advantage. You had the three things needed to make the lever work:

1. The load—your friend on one end.
2. The fulcrum—the center support of the seesaw.
3. The effort—you pushing down on the opposite end.

A seesaw is a lever with the fulcrum between the load and the effort.

Make a First-class Lever

A seesaw is a first-class lever. A first-class lever has the fulcrum between the effort and the load. Experiment with your own first-class lever.

You will need:
- rigid ruler
- table
- small bean bag (you can make a bean bag by filling a small sock with dried beans or rice)

Place the ruler so half is on the table and the other half hangs over the edge. The table edge is the fulcrum. Place the bean bag (the load) on the end of the ruler that rests on the table. Now, press down (the effort) on the opposite end of the ruler and lift the load. Notice how far the bean bag was raised. Notice how much effort you used to lift the load.

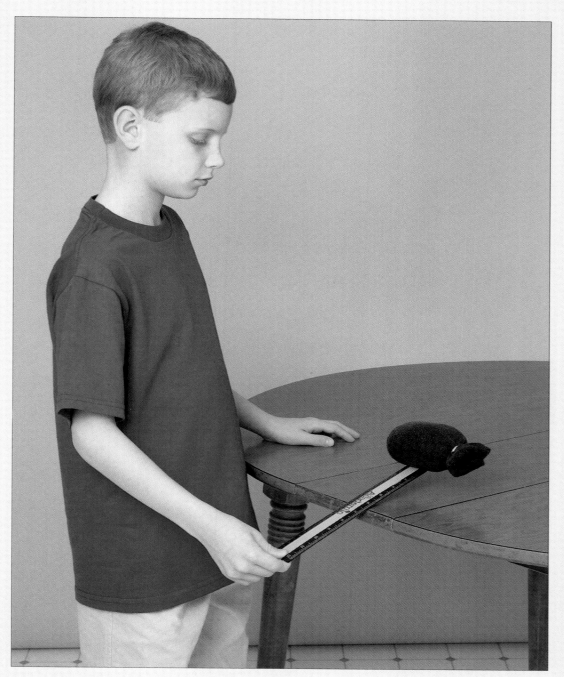

You can use a table edge, a ruler, and a bean bag to do a simple lever experiment.

Changing Force and Distance

Do the first-class experiment again. This time place the ruler so that only 2 inches (5 centimeters) hangs over the edge of the table. Place the bean bag on the end of the ruler that rests on the table, the same way you did in the first experiment. Press down on the opposite end of the ruler and lift the load. How far did the bean bag rise from the table? Did you use more or less effort to lift the load?

Moving the ruler will change the position of the fulcrum.

Weight Arm and Force Arm

The position of the fulcrum is important when you are using levers. In a first-class lever, the part of the lever between the fulcrum and the load is called the **weight arm**. The part between the fulcrum and the effort is the **force arm**.

In the last experiment, the fulcrum was closer to the effort than to the load. In other words, the force arm was shorter than the weight arm. When you have a shorter force arm than weight arm, more effort is needed to lift the load.

A giant first-class lever makes it possible for a boy to lift a car.

Make a Second-class Lever

A second-class lever has the load between the effort and the fulcrum. A wheelbarrow is a second-class lever. Think about what happens when you grab the handles of a wheelbarrow and move it across the yard. Where are the effort, the load, and the fulcrum?

You will need:

- 3 feet (90 cm) of string
- paperback book
- friend
- rigid yardstick
- table
- masking tape

To prepare the load, tie one end of the string around the book. To prepare the fulcrum, have your friend place about 2 inches (5 cm) of the yardstick on the table's edge.

The remainder of the yardstick should extend away from the table. Tape the 2 inches (5 cm) of the yardstick to the table. The yardstick is your lever. Your lever should be able to move up and down while the taped section works as the fulcrum.

Use a yardstick to make a second-class lever.

Use Your Second-class Lever

With the load on the floor, grab the free end of the string and lift the load about 6 inches (15 cm) from the floor. Notice the amount of effort you used.

Now, place the book on the floor centered under your lever. Tie the free end of the string around the lever leaving no slack in the string. Holding the free end of the lever, lift the load about 6 inches (15 cm) from the floor. Notice the amount of effort you used. Did the lever make your work easier?

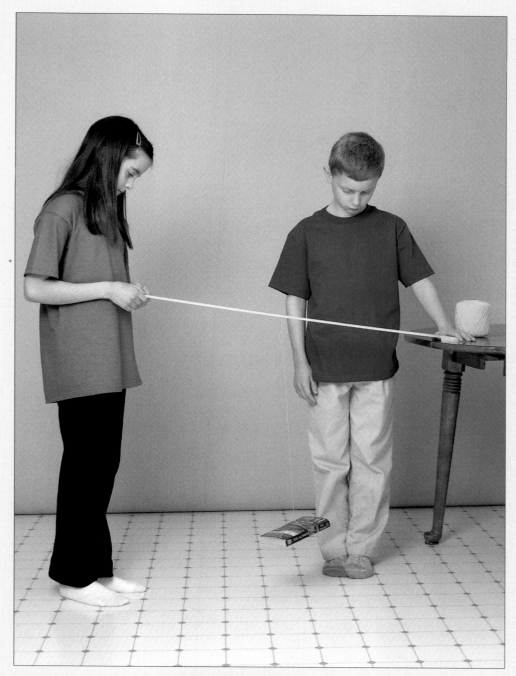

It takes less effort to lift a load using a lever.

Make a Third-class Lever

You can make a slight change to the last experiment and make a third-class lever. A third-class lever has the effort between the fulcrum and the load.

Move the string to the free end of the lever, placing the load as far away from the fulcrum as possible. Stand between the fulcrum and the load (the effort in the middle) and lift the load. This type of lever uses the principle of a third-class lever. Can you think of any household objects that are third-class levers?

A broom is a third-class lever. One hand steadies the broom handle at the top (fulcrum). The other hand grasps the broom handle in the middle and sweeps (effort). The dirt on the floor (load) gets collected into a pile.

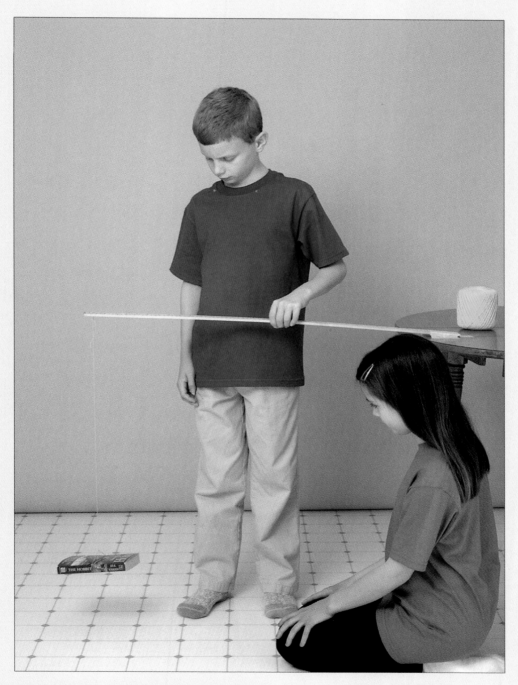

A third-class lever has the effort between the fulcrum and the load.

Common Levers

People use levers every day. Can you spot any levers around the house or classroom? Scissors, nutcrackers, and tweezers are all double levers. Double levers are two levers joined by a fulcrum.

A pair of scissors is a first-class double lever. The fulcrum lies between the effort and the load.

Nutcrackers are second-class double levers. The load is between the effort and the fulcrum. Tweezers are third-class double levers. The effort is between the load and the fulcrum.

When you use a nutcracker, you place the load between the effort and the fulcrum.

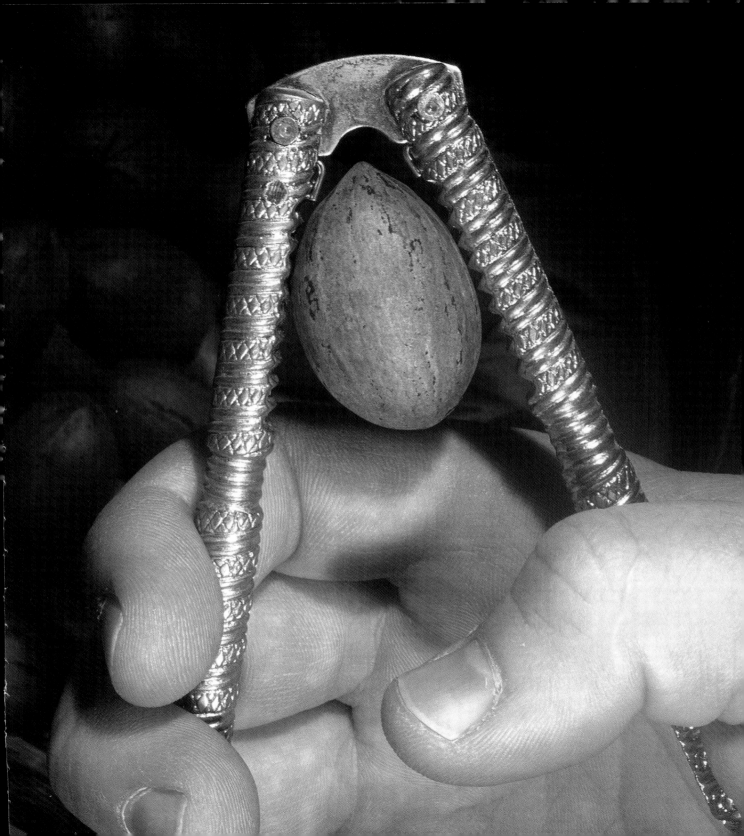

Index

first-class levers   16, 21, 28
force arm   21
second-class levers   22, 24, 28
third-class levers   26, 28
weight arm   21

About the Authors

David and Patricia Armentrout have written many nonfiction books for young readers. They specialize in science and social studies topics. They have had several books published for primary school reading. The Armentrouts live in Cincinnati, Ohio, with their two children.